A Beginning-to-Read Book

EARTH SCIENCE

WEATHER NEAR YOU

by Mary Lindeen

NORWOOD HOUSE PRESS

DEAR CAREGIVER, The *Beginning to Read—Read and Discover Science* books provide young readers the opportunity to learn about scientific concepts while simultaneously building early reading skills. Each title corresponds to three of the key domains within the Next Generation Science Standards (NGSS): physical sciences, life sciences, and earth and space sciences.

The NGSS include standards that are comprised of three dimensions: Cross-cutting Concepts, Science and Engineering Practices, and Disciplinary Core Ideas. The texts within the *Read and Discover Science* series focus primarily upon the Disciplinary Core Ideas and Cross-cutting Concepts—helping readers view their world through a scientific lens. They pique a young reader's curiosity and encourage them to inquire and explore. The Connecting Concepts section at the back of each book offers resources to continue that exploration. The reinforcement activities at the back of the book support Science and Engineering Practices—to understand how scientists investigate phenomena in that world.

These easy-to-read informational texts make the scientific concepts accessible to young readers and prompt them to consider the role of science in their world. On one hand, these titles can develop background knowledge for exploring new topics. Alternately, they can be used to investigate, explain, and expand the findings of one's own inquiry. As you read with your child, encourage her or him to "observe"—taking notice of the images and information to formulate both questions and responses about what, how, and why something is happening.

Above all, the most important part of the reading experience is to have fun and enjoy it!

Sincerely,

Shannon Cannon

Shannon Cannon, Ph.D.
Literacy Consultant

Norwood House Press • Chicago, Illinois
For more information about Norwood House Press please visit our website at
www.norwoodhousepress.com or call 866-565-2900.
© 2018 Norwood House Press. Beginning-to-Read™ is a trademark of Norwood House Press.
All rights reserved. No part of this book may be reproduced or utilized in any form or by any
means without written permission from the publisher.

Editor: Judy Kentor Schmauss
Designer: Lindaanne Donohoe

Photo Credits:

All photos by Shutterstock

Library of Congress Cataloging-in-Publication Data
    Names: Lindeen, Mary, author.
    Title: Weather near you / by Mary Lindeen.
    Other titles: Beginning-to-read book.
    Description: Chicago, IL : Norwood House Press, [2017] | Series: A beginning
    to read book | Audience: K to grade 3.
    Identifiers: LCCN 2017002618 (print) | LCCN 2017005797 (ebook) | ISBN
    9781599538730 (library edition : alk. paper) | ISBN 9781684041053 (eBook)
    Subjects: LCSH: Weather–Juvenile literature. | Weather forecasting–Juvenile literature.
    Classification: LCC QC981.3 .L55 2017 (print) | LCC QC981.3 (ebook) | DDC
    551.5–dc23
    LC record available at https://lccn.loc.gov/2017002618

Library ISBN: 978-1-59953-873-0   Paperback ISBN: 978-1-68404-092-6

352R—052022
Manufactured in the United States of America in North Mankato, Minnesota.

What's the weather like today where you live?

It's a warm, sunny day here.

The sky is blue.

The sun is shining.

It's a rainy day here.

The sky is cloudy.

Everything is wet.

It's a snowy night here.

It's cold outside.

Everything is blowing around!

It's a windy day here.

You can find out about the weather by looking out the window.

You can find out about
the weather by
looking it up
on a computer.

And you can find out about the weather by watching the news.

WEATHE[R]

	MON	TUE	
	80°	82°	80°
	75°	72°	76°

ORECAST

ED	THU	FRI	SAT
1°	79°	80°	82°
5°	79°	70°	72°

Did You Know?

A meteorologist is a person who studies weather patterns, tracks how they change, and tells people how to stay safe in dangerous weather.

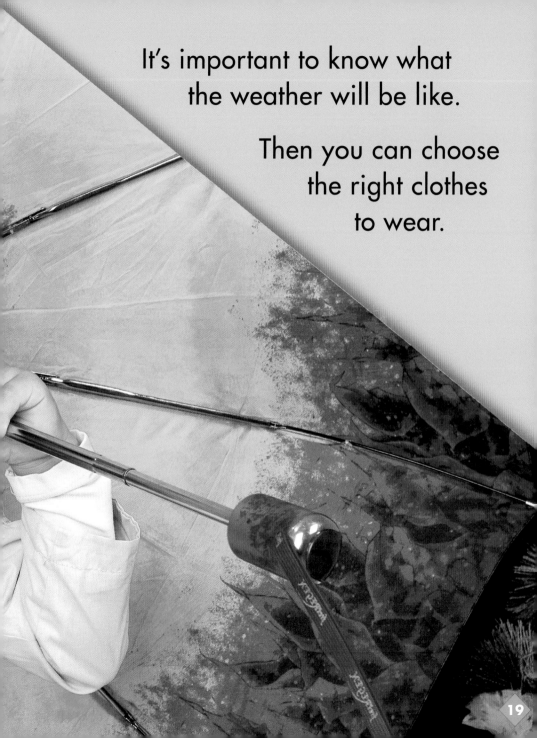

It's important to know what
the weather will be like.

Then you can choose
the right clothes
to wear.

Paying attention to the weather
is also important because
it can change.

Did You Know?

Every day there are about
1,800 thunderstorms happening
somewhere on Earth.

A lot of rain can become a storm.

A lot of wind can become
a tornado.

A lot of snow can become
a blizzard.

So before you go outside, be sure to check the weather!

sunny

rainy

snowy

windy

storm

tornado

blizzard

CONNECTING CONCEPTS

UNDERSTANDING SCIENCE CONCEPTS

To check your child's understanding of the information in this book, recreate the following graphic organizer on a sheet of paper. Help your child complete the organizer by identifying some of the characteristics of different kinds of weather.

Sunny	• • •
Rainy	• • •
Snowy	• • •
Windy	• • •

SCIENCE IN THE REAL WORLD

Draw a 7-column chart on a piece of paper. Have your child label the columns with the days of the week. For the next seven days, keep track of the weather and write it on the chart. Draw a picture of something fun you did on each of those days. Talk about how the weather influenced what you did each day.

SCIENCE AND ACADEMIC LANGUAGE

Make sure your child understands the meaning of the following words:

weather	computer	news	meteorologist	patterns
dangerous	thunderstorms	tornado	blizzard	

Have him or her use the words in a sentence.

FLUENCY

Help your child practice fluency by using one or more of the following activities:

1. Reread the book to your child at least two times while he or she uses a finger to track each word as it is read.

2. Read a line of the book, then reread it as your child reads along with you.

3. Ask your child to go back through the book and read the words he or she knows.

4. Have your child practice reading the book several times to improve accuracy, rate, and expression.

FOR FURTHER INFORMATION

Books:

Boothroyd, Jennifer. *What Is Today's Weather?* Minneapolis, MN: Lerner, 2014.

Jackson, Tom. *Magic School Bus Presents: Wild Weather*. New York, NY: Scholastic, 2014.

Rattini, Kristin Baird. *Weather*. Washington, DC: National Geographic, 2013.

Websites:

For Kids Network: Weather for Kids
http://www.weatherforkids.org/

Web Weather for Kids
http://eo.ucar.edu/webweather/

National Geographic for Kids: 30 Freaky Facts About Weather
http://www.ngkids.co.uk/science-and-nature/30-freaky-facts-about-weather

Weather Near You uses the 92 words listed below. *High-frequency* words are those words that are used most often in the English language. They are sometimes referred to as sight words because children need to learn to recognize them automatically when they read. *Content words* are any words specific to a particular topic. Regular practice reading these words will enhance your child's ability to read with greater fluency and comprehension.

High-Frequency Words

a	because	find	know	right	to
about	before	go	like	so	up
also	blue	here	look(ing)	tell(s)	what
and	by	how	of	the	where
are	can	in	on	then	who
around	day	is	out	there	will
be	every	it	people	they	you

Content Words

attention	cold	lot	safe	sunny	weather
become	computer	meteorologist	shining	sure	wet
blizzard	dangerous	news	sky	thunderstorms	what's
blowing	Earth	night	snow(y)	today	wind(y)
change	everything	outside	somewhere	tornado	window
check	happening	patterns	stay	tracks	
choose	important	paying	storm	warm	
clothes	it's	person	studies	watching	
cloudy	live	rain(y)	sun	wear	

About the Author

Mary Lindeen is a writer, editor, parent, and former elementary school teacher. She has written more than 100 books for children and edited many more. She specializes in early literacy instruction and books for young readers, especially nonfiction.

About the Advisor

Dr. Shannon Cannon is an elementary school teacher in Sacramento, California. She has served as a teacher educator in the School of Education at UC Davis, where she also earned her Ph.D. in Language, Literacy, and Culture. As a member of the clinical faculty, she supervised pre-service teachers and taught elementary methods courses in reading, effective teaching, and teacher action research.